THIS BOOK BELONGS TO :